I0755759

PROPHECY

THE ANSWER TO

ISLAMIC JIHAD

AND

GLOBAL WARMING

T. M. PALMER

Printed in the United States of America

Published by All Answers Press,

Post Office Box 2666,

Rancho Santa Fe, California 92067

U.S.A.

Library of Congress Cataloging in Publication Data

ISBN 978-0-6151-8226-1

This book is dedicated to the continuation of Western Civilization.

PROPHECY

THE ANSWER TO

ISLAMIC JIHAD

AND

GLOBAL WARMING

Chapter One

Atomic explosions have devastated Portland, Oregon and Boston Massachusetts within forty minutes of each other resulting in the deaths and injuries of over one million Americans. Al-Qaeda, through the news organization, Al-Jezeera, has demanded the surrender of the United States of America and submission to the rule of an Islamic Religious Council

Otherwise more cities will be destroyed. When this day comes, how do you want the President to respond?

The above scenario is at the heart of the conflict between America and the Nation of Islam. As such, a variation of the description will appear at the beginning of each of the following chapters.

The reader may notice that terror experts and even the highest ranking military and government authorities readily assert that atomic weapons being used by Muslims against America is not only a real risk but that it appears increasingly that it is virtually inevitable. What is not presented by these authorities is the accompanying demand for surrender. While the above scenario is understood by all in positions to know, since we have neither a deterrent nor an answer to the dilemma, it is simply not dealt with or even brought to the public's attention. Thus, the threat is always portrayed as an isolated terror event, like the attack on the World

Trade Center on 9/11, one to be suffered and endured.

The reality is that with only a few nuclear detonations, America may well be brought to her knees in surrender. This is the real threat that is never brought to your attention. Currently we, as a nation and a people, have no deterrent or adequate response except further devastation or surrender.

Consider for a moment what the above threat actually means. If the President refuses to surrender after two cities are destroyed…possibly because of a desire to believe that the Muslims have no more nuclear weapons in place to destroy other American cities…and then two more cities, then two more are destroyed, then what does the President do?

If after a few days six major American cities are reduced to radioactive scenes of destruction and death, probably large numbers of people will begin fleeing the cities. This evacuation may be ordered or spontaneous but the result will be the same. American society and economy, based upon the efficiency of cities, cannot support a hundred million refugees evacuating into the rural areas. There will be no shelter, water, food, fuel, medical care, or any other necessary goods and services with all the distribution systems inoperable. All the shipping, rail, and truck systems for moving the vast amounts of goods will not be sustainable and the result will be millions dying from a lack of these necessities that modern societies depend upon for survival. In comparison, the Katrina incident would be nothing. What then should the President do?

The biggest threat to America since World War II and the Cold War is the current war with Islam and the Islamic populations, principally the Arab, Iranian and Indonesian populations.

Why are we at war with these people? We are being attacked because the Muslims have decided, from the teachings of their religious leaders, that we, as infidels, are as despicable vermin, not worthy of life outside of being slaves, and all good Muslims have a religious duty to destroy, to kill us. They have chosen to believe, just as the German Nazis and the Imperial Japanese did before and during World War II, that they are superior to all others who are not as they are.

There is in practicality, no reason to be concerned with specific passages in the Koran regarding the killing of infidels. While these direct commandments are there and have been often quoted in discussions of this holy war, what really matters is the state of mind of the Muslims. They are led not only by the direct words in the Koran but more importantly by the clerics and mullahs who are influencing the people.

There are really no contradictions in the Koran, rather, what appear to be conflicts are a change in the message from Allah in the over two decade-long "lesson", or revelation, to Mohammad brought by the Archangel Gabriel. Unlike the Holy Bible which is a collection of works by different authors, the Koran is supposedly the direct word of God (Allah) to Gabriel and who then tells it, in Arabic, to Mohammad who then writes it down, dictates it to others to write, or simply remembers it to dictate later. The "contradictions" are really changes as time goes by. In particular, in the first lessons, Mohammad is advised that many people will not heed him or will pretend to believe and not be sincere. Mohammad is to take this in stride and not be discouraged. In particular there is the passage that demands acceptance and respect of Jews and Christians because they too are monotheists and accept "Judgment Day".

It is common for Westerners to take refuge in certain passages where the Koran proclaims that Allah is merciful and that Muslims should be merciful. However, as the Koran progresses over the years it becomes plain that Muslims should be merciful only to other Muslims. All others should be killed and that it the message being taught today throughout the Nation of Islam.

It is most important to understand that Muslims, from Mohammad on, want to make the world Islamic. To spread and defend Islam is the goal and everything is allowed, and nothing is disallowed, to accomplish this. In particular, lying and deceiving the Infidel is allowed. This is why we are so often being told by Muslims that quote passages from the first few lessons that Muslims will honor and accept and live in peace with other religions while they know and understand what they are really doing is deceiving... It should always be remembered that when Mohammad attacked Medina, which was predominantly Jewish and Christian during his time, he personally slaughtered Jews and Christians with his own sword, cutting of the heads of what were considered infidels.

That an entire culture of people would be liars is hard to accept for most in the West. We want to think that all people share our idea of truth, values, sense of morality, virtue, and fair play. We want to think what we believe is self-evident to all others. This is very naïve, to say the least. To accept the great differences between people of different cultures, perhaps the following will help.

While Muslims are most certainly not cannibals, it might be helpful to consider how a cannibal would regard things in everyday life differently than we would. It is historical fact that humanity has produced numerous cannibal societies over the millennium.

Imagine if a cannibal from Borneo or Papua New Guinea moved into your neighborhood. Would not such a person, sharing love for his family, hard working, imaginative, still not regard the nine year old girl across the street differently than we would consider appropriate? As he would think of her as tender, tasty, succulent, comparing her to someone older and determining the difference in tenderness and flavor, as he would think in terms of how to capture, butcher, skin, dismember and cook her be a different way of regarding her? As he would carry out the above actions, would he not be considered virtuous? As he brings home meat to his family's table, supporting his family, clan, village, and nation, is he not doing God's will? Is he not, in his and his peoples' eyes anything other than moral, industrious and even noble?

As the cannibal would look at his relationship with that little girl in a manner that is completely one sided, with all benefit accruing to only one and all disadvantage, even unto death, borne by the other party, so too do Muslims regard the Non-believer as worthy only for death. Thus it is with the way the Muslim regards the Infidel.

As much as we might want to believe that Muslims are the same as us, they believe in a very different sense of morality. Even today, Muslim clerics verify that Islam condones slavery. Indeed, while black Americans were striving to achieve complete civil rights in America in the late fifties and early sixties, slavery was alive and well in the heart of Islam. Slavery was widespread in Saudi Arabia until 1960 and not outlawed until 1962 under immense pressure from Western countries. Today Sudan is still a stronghold of slavery not unlike the way it has been practiced in Africa for centuries.

It is imperative that we in the West accept that the Muslims will overtly lie to us. They are culturally required to lie to defend family, clan, tribe, nation, and most importantly, Islam. The infidel is not even remotely deserving of the truth if it in any way puts Islam or Muslims in any bad light or makes them in any way less. This is why we will find ourselves under increasing and constant pressure to change our way of doing things in favor of the Muslim way and they never agree to change. An excellent example of this is the total ban of religion in U.S. public schools while at the same time allowing Muslim students to pray five times a day, complete with religious leaders and prayer rugs not only being allowed but required of the schools.

Later, the Koran becomes more forceful of killing all "non-believers" in the oneness of Mohammad as the only true Prophet of Allah. Many Koran scholars, clerics and mullahs have rendered different interpretations as to what an infidel is. Many Sunnis think all Shiites are infidels and should be killed. Some believe that a Muslim who does not pray five times a day is an infidel and should be killed.

Of course the most telling characteristic is that a Muslim cannot leave Islam upon pain of death. This presented an enlightening example in 2006 when an Afghani who had left his country some years before for Italy and converted to Christianity. Upon returning to his homeland it was proclaimed by all the clerics and mullahs that he must be put to death for renouncing Islam, which threatens Islam. This death demand came not from a few isolated extremists from the Taliban, but from the ordinary clerics and Muslim people. This presented an enormous problem for President Karzai of Afghanistan who enjoyed thousands of Christian American troops fighting and dying for Afghan freedom from Islamists. The only solution they could arrange was to declare the apostate insane and ship him back to Italy in order to keep his alive.

The Bible has passages too that one would ignore in a practical manner, like the actual orders from God in Genesis where God specifies the correct punishment of slaves. God stipulates when to cut out an eye, when to kill the slave, etc. Never does God in the Old or New Testaments, the Torah, or the Koran, ever say not to enslave. One could easily conclude that God endorses and condones slavery, but this is not relevant today except in some Islamic circles. Similarly, in Genesis God commands that children be stoned to death for being disrespectful to their parents. We don't kill our kids for mouthing off at us.

However, irrespective of the exact verbiage in the Koran, Muslims do kill out of religious duty and supposed command of Allah. They actually do these things. A Palestinian father, upon finding out his daughter blew herself up as a suicide bomber in Israel, was filled with righteous rage and stated he wished he could strangle her with his own hands. This was not over her killing other people but for shaming him by leaving the house without his permission.

Death and sex are the two main driving characteristics of Islam. Muslims want to kill for honor, saying any thing against Islam, women leaving home without the permission of the head male of the household, women having sexual affairs or even talking to men outside the home, failing to pray five times a day and so on. Muslims are obsessed with sex as are few others. They want their women shrouded and veiled. A man can have four wives; can divorce a wife just by saying "I divorce you" three times. A woman cannot prove rape without four male witnesses to the crime. The list goes on and on. There was a case in Iran in 2006 where an engaged couple were walking together in a park. Several Islamist "religious police" beat and killed both of them for being together unmarried. The Islamic court ruled the act of killing them was justified.

There is no other significant religion, not Confucianism, Sikhism, Hinduism, Judaism, Buddhism, or any of the Christian religions, that has as a basic tenet of belief, that the adherents should kill all non-believers. Almost everyone thinks they are better because they know God in a superior way but no others want to kill everybody else. The acceptance of Islam is quite different from respecting Islam which will, in the end, provide the answer to this problem.

Chapter Two

Atomic explosions have devastated Chicago, Illinois
and Atlanta, Georgia within forty
minutes of each other resulting in the deaths
and injuries of over one million Americans.
Al-Qaeda, through the news organization,
Al-Jezeera, has demanded the surrender of the
United States of America and
submission to the rule of an Islamic Religious
Council
Otherwise more cities will be destroyed. When
this day comes, how do you want the President
to respond?

Why now? What has brought about this current relentless rise in conflict and death? The answer is MONEY! For centuries the Arabs have been essentially poor and not a threat to anyone but themselves with their constant warring amongst each other. The addition of vast wealth from oil gave the Arabs, Iranians and Indonesians the power to revive their long held desire to conquer the world for Islam. Before they had the money, they simply were not in a position to wage war. Money was needed by Alexander, Charlemagne, Caesar, Genghis Khan, Napoleon, Hitler and every other conqueror in history. The more money; the more men and arms produce the greater pride and avarice. Just as the Spanish enslaved and converted the Americas, fueled with Aztec and Inca gold, the oil riches are fueling the current war of Islam against the West, most notably by the Saudis and their disciples of the extreme form of Islamic teachings emanating from the city of Wahabi, in Saudi Arabia. It is this brand of Islam that Osama Bin Laden, rich from money paid to him and his family by the Saudi Royals, adherers to to this day. Along with many rich and even royal Arabs from the various Arab Persian Gulf countries, he has been the focus behind the attacks on America and other Western countries. Pakistan is the most dangerous for it is nuclear while Iran is the next most dangerous since it alone is the most powerful and rich theocracy and they are going nuclear.

This war has been waged against us by the Nation of Islam for several decades now. The Western World has largely ignored it and treated it as a minor irritation being far less important than acquiring oil and penetrating the Islamic populations for commercial profit. Especially under the deeply religious presidencies of the Bush and Carter administrations and the "out to lunch" Clinton administration, we have approached the growing menace with a very benign strategy of containment and cultural modification. The George W. Bush Presidency in particular has concluded that we can defuse the threat by bringing democracy to these populations, no matter with how much chaos.

This is wrong for several reasons, not the least of which is that the Muslim insurgency thrives on chaos since it is a true grass roots movement.

What Muslims will care about is not being killed by other Muslims, but they really have no reluctance at having Americans killed by Muslims. We can see that the people, given a true democratic vote, as in Iran and in the Palestine Authority, clearly approve of anti-American governments who foster holy war against the West.

While people in the West might still be trying to figure out Hamas, clearly the people in Palestine knew exactly what they voted for. We are being fed save-face propaganda that Hamas won the vote because they were better organized and had good civil programs and were less corrupt. The people know Hamas wants holy war and that's why they voted for them. Then, of course, there are all the Fatah elements that want war just as much but simply want their party to be in power during the war. A huge majority of the people want holy war. They want to dominate the world for Islam, as is their religious duty. It would be most unwise to forget or ignore the pictures of vast numbers of

Palestinians who danced joyously in the streets at the news of thousands killed on 9/11.

Realistically, democracy is for more advanced cultures. The ancient Greek city-states, from which we draw our ideas of democracy benefited from a sophisticated progression that over hundreds of years slowly developed into a viable form of government. Rome, over two thousand years ago, also used Greek ideas and forms of political organization to form another successful democracy. Over time, both Greek and Roman democracies suffered a decline into dictatorship and ultimate defeat.

Christian culture suffered for centuries in the medieval "dark ages" and was utterly unable to create democracies until a vast improvement in collective thinking allowed the Renaissance to flower, leading to the European "Age of Reason" which produced our American democracy.

The Nation of Islam is still very much a medieval culture and is not ready for democracy and will not be for centuries as they are now going. Their view on democracy is to destroy it and return the world to medievalism

The people of Afghanistan, happy to be free from the extreme form of Islam under the Taliban, were happy to participate in democracy, proudly showing off their purple fingers proving they had voted. As the Taliban has reemerged as a power, the Afghans are forced to rethink their choice and many are likely to prefer the relative peace rather than the continued war. They have little to no commitment to democracy.

President George W. Bush has stated that "Democracies don't attack other countries." He should read American history and an incomplete list of military incursions would list: Canada, Russia, innumerable Indian nations, Vietnam, Laos, Cambodia, Egypt, Cuba, Mexico, Panama, Granada, Morocco, Libya, Tunisia, Philippines, Kingdom of Hawaii, Guam, China, not mentioning nations we were at truly legitimate war with like Germany, Italy, and Japan., not to leave out history's two greatest democracies, Athens and Rome, both of which lived to conquer other people.

He could not be more wrong than in thinking he can bring peace, democracy and prosperity to Muslim countries through democracy until we render them ready for it, as we did with Japan.

Then there is truly the most important question: Do we Americans owe those who wish to kill us our blood, treasure, peace and likely even our own culture, nation and freedom? The Muslims of today, at least the ones that have the spiritual energy to move events, are the ideological successors of the Medieval Christians of fifteen hundred years ago. Human nature being what it is, responds to different influences but excitement and religious fervor have always fueled an increased spiritual energy. Pope John Paul II sadly admitted the Muslims had more spirit than Christians. A call to war is always more exciting than a call to peace. No one gets excited about living a quiet, calm and prosperous life when the alternative is rage, excitement, hatred, religious or political fervor whipped up to hysterical frenzy by demigods. The Muslims are on a rampage and everything from Papal remarks to cartoons are enough to unleash their rage to the extent that now no one dares to speak out about Islam as they would about anything else. This is true terrorism.

When we want to understand the Nation of Islam, all we have to do is understand our own Christian forebears during the dark ages. As Europe was a combination of kingdoms, baronies, and other greater and lesser feudal powers, they were all under the greater power of the Roman Catholic Church

There were really two forms of government, the secular control asserted by kings, barons, counts, etc over their individual fiefdoms and the control exercised by Rome, the Pope, and the various Cardinals, Bishops, and other clergy. With the power to excommunicate, the Pope could control the kings. The people were serfs and belonged to the land but they were first Christians and loyal to the Pope, the Vicar of Christ and who as such connected the people to God. The Pope could declare holy war as was done with the 300 year Crusades. Today, the Nation of Islam operates in much the same medieval manner.

When the Pope sent forth the Crusades to battle the Nation of Islam and win back Jerusalem, the religious power superseded the secular power and so it is today in the Nation of Islam. We want to deal with the secular governments and this is a waste of energy as the power that is the enemy is the religious power wielded by the clerics and the mullahs and Ayatollahs. This is the power most Muslims obey.

The Muslims who were attacked in the Crusades were not being attacked, for example, by the King of France in his secular capacity, but by the King in his capacity as a Christian ruler who commanded French Christians.

The Christians were fighting not only to recover Jerusalem but to protect Europe from Muslim incursions and growing influence. This is also a fear of the Muslims today and some cite this as a reason for Jihad.

We are at war with the Muslim people, who are actually being led in this Holy War by the clerics. This fact is inescapable. We have to accept this and act accordingly. It is wise to remember that negotiating with those who are doing God's Will is futile. The Muslims' idea of turning the other cheek is that of the infidel's severed head rolling on the ground.

Chapter Three

Atomic explosions have devastated
Dallas, Texas and Seattle, Washington
within forty minutes of each other resulting
in the deaths and injuries of over
one million Americans.
Al-Qaeda, through the news organization,
Al-Jezeera, has demanded the surrender of the
United States of America and
submission to the rule of an Islamic Religious
Council
Otherwise more cities will be destroyed. When
this day comes, how do you want the President
to respond?

We have the ability to defend ourselves in this war if we are willing to do so. Unfortunately, so far we have allowed ourselves to regard this threat in a manner that may doom America to failure and defeat. Specifically, we are doing two very wrong things. First, we are not appreciating how much danger we are in and second, we are adhering to a method of fighting that very recent history has already taught us fails miserably.

History provides us with numerous examples of how a very large number of people are defeated by a small number of determined and cunning adversaries. Whether it is the destruction of the great city-state of Troy with a large wooden horse concealing a small number of warriors or the holocaust of the European Jews at the hands of the Nazis such acts are often repeated. Close to home in our own hemisphere is a very telling example of how this can happen.

In 1532 Francisco Pizzaro the Spanish conquistador, with only about 250 soldiers, (not counting the priests, alter boys, cooks, etc.) and employing duplicity and audacious murder, was able to utterly defeat a great civilization of about fifteen million people, the Inca of Peru. Then, as now with this war, religion played a pivotal role.

We are collectively close to the mind set of the German Jews in the early 1930's when the Nazi attacks against them were intensifying. The Jews, and the rest of the world, simply could not imagine, could not conceive, the fate that lay before them. The thought of the brick through a window or the beating of a Jewish neighbor as a matter for law enforcement to deal with was common. The trains of death, the gas chambers, the ovens, the stacks of their hair, the piles of gold from their teeth, the murder of millions was simply not within their frame of reference. So it is with us.

When it comes to the issue of religious holy war our country and cultural beliefs act strongly upon us to render us as vulnerable as were the Jews. We have such an ingrained belief in religious freedom that we have an almost impossible capacity to understand when we are in a holy war. Our frame of reference is that no matter what the religion is we must accept it, honor it, and allow its free expression, and we can't fathom the possibility that it exists to destroy us.

One of the interesting things which characterize the current conflict between the Muslims and the West is that we are not, for the most part, fighting nations. We are used to having a population be identified by their nationality. This is not applicable currently for two main reasons.

First, although all Arabs and the Iranians are citizens of a nation, be it Egypt, Jordan, Sudan, Yemen, Saudi Arabia, etc, those people are not truly represented by their governments which are almost entirely dictatorships. (Iran is probably the exception here for the current radical Islamist government was chosen by the Iranian people who are clear in their popular desire for both possession of atomic weapons and the globalization of Islam.) The people are far more influenced by the mullahs and the teachings in their mosques. The people are really quite removed from their governments and this is why on the one hand we hear conciliatory comments and even some actual assistance from the governments on our "war on terror", yet on the other hand, it is the people of these countries who are truly the ones who are the enemy.

We insist that if we are at war it must be against a sovereign nation, a Germany or a Japan. We were similarly unable to fight against widespread warfare that entailed armies fighting us from areas that we couldn't have a clear idea of nationhood, as in Cambodia, Vietnam, Somalia, and a few others We think we can only wage war on a clearly defined government, not a people. We reject the idea of being at war with a population. We think only governments are wrong while the people are always nice and deserving of our saving them.

The second major mistake America is making is to cling to the failed "Hearts and Minds" strategy that was dreamed up during the Vietnam War and which proved so utterly stupid. Somehow our leaders came out of the victory of World War II which produced the complete defeat of our enemies and was then followed by the effort to rebuild under the Marshal Plan (and similar efforts in Japan) completely confused about why that worked so well.

After World War I, with Germany destroyed (along with much of other regions of Western Europe) the Germans suffered perhaps the worst during the global economic depression of the early 1930's which was so hard even here in America. Hitler tapped into the anger and feelings of deprivation in the minds of the German people and used this as a means of rallying them to rebuild their war machine in spite of the restrictions of the Versailles Treaty which Germany signed at the end of World War I. Failure to enforce the treaty led to the rise of German power and resulted in the deaths of tens of millions in the years ahead.

It must be recognized that history teaches us an absolutely true fact: major wars are the result of inaction early in a time of rising war risk. If France and England had acted forcibly against Hitler early on before he could rebuild his military tens of millions of lives would have been saved. It is the lack of the capacity to tolerate a low level of destruction in order to achieve a greater good that condemns us to massive death and destruction later.

Winston Churchill understood clearly this reality in the months leading up to WWII. When the appeasement of Hitler was the real accomplishment of Neville Chamberlain, then Britain's Prime Minister, Churchill, who had called for force to be used against the Germans, declared, "You wanted honor and no war. You will have war and no honor!" How right Churchill was.

The dominance of political correctness in American culture is the liberal religion. Political correctness has become our version of honor. This idea is hobbling us and destroying our ability to act in our own self interest. In fact under this concept we are the only people on earth who should not have self interests at all. This takes on a concrete application when we have to act in self defense.

Intent on not repeating the mistakes of post WWI, American and British forces made an amazingly successful effort to rebuild Germany rather than allow it to languish in chaotic ruin as had happened before. The key difference is that the Germans and Japanese were utterly defeated: militarily, spiritually, economically, and socially. It was only after these passages through the fire of defeat were they ready to respond positively to a new reality. There was even a change in religion in that the Japanese had to end their centuries-old belief that the emperor was a god. (Even in Germany, Hitler was essentially backing a religious war. They were Christians waging war against the lesser mortals: Jews, Gypsies, homosexuals... Hitler even wanted an alliance with America to allow us to jointly fight and kill the Jews. Hitler did enjoy the Grand Mufti of Jerusalem's desire to help him exterminate the Jews...all of them.)

This laudable example following WWII was brought into play again after the invasion by the North Korean communists and the partitioning of Korea. Massive amounts of capital poured into South Korea and have helped make that country one of Asia's most prosperous and democratic nations, while still threatened by the North Koreans.

Then again, in another civil war, this time in Vietnam, another effort was tried to employ the same agenda. Here the end was quite different than in Korea. Largely because of the defense pact between the U.S.S.R. and North Vietnam, we did not wage effective war and entered into a war to win the "Hearts and Minds" of the Vietnamese people. As history has shown, this was a horrible failure that resulted in the deaths of over 50,000 American service personnel. The key point of the war was that we Americans were not there as conquerors but as an army trying to keep the people of South Vietnam free from communism.

Unfortunately for the Vietnamese, the energy and the commitment on the part of the communists proved to be the winning side. Of particular note, in both the Korean and Vietnam Wars, never did the people of those two countries ever attack the US. We went there to try to contain communism and help them and in the process produced the deaths of hundreds of thousands. We had one decent draw and one total defeat.

From the 1950's to the 1990's our potential conflict with the Soviet Union took several turns but stayed essentially a "Cold War". We never came to armed conflict with the Soviets although at different times it came close. We were able to prevent a communist invasion of Greece and we turned away the missile threat in Cuba but we ended up in Vietnam. The Soviet Union finally fell apart from internal weakness, aided in large part by the Chernobyl disaster which undermined the faith of the Soviet Eastern Bloc in Moscow's leadership and by the anti-communism vote in Nicaragua which was the final blow to European faith in communism itself. Not to forget Afghanistan. Of course the principal reason the Soviets never attacked us was because there was absolutely no doubt in anyone's mind what would happen to them if they did...we would annihilate them by the hundreds of millions.

It was this totally credible deterrence that kept us safe from the Soviets, in spite of the propagandized hatred in the minds of millions of Soviet military people and the Soviet citizens. One thing they all knew was that they could not tolerate anyone taking the matter into their own hands and launching an attack on America. It became the responsibility of the Soviet people to maintain control over their own upon pain of death for everyone.

Undoubtedly many millions of Communists believed in their near religious fervor of the ultimate rightness of communism over the wicked capitalist West. Marxism-Leninism propaganda was refined to a basic position: for everyone to be prosperous everyone had to be communist. The refusal of the West to submit to communism meant we were the ones causing all of the misery in communist societies and that gave them the right to kill us as "self-preservation".

Essentially, the proponents of Islam believe they too have to turn the world to Islam and they have the God given right to kill all infidels who refuse to believe in Islam and submit to theocracy.

It is only a matter of time before the Islamists find a way to detonate atomic weapons and kill million of Americans. This view is held by virtually all experts in the field of defense. Now we come back to the present war which is truly World War III. This time it is really global in its reach and we have been targeted and attacked on our own soil for the first time since the War of 1812. (The attack on Pearl Harbor was not on U.S soil since the Kingdom of Hawaii did not become a state until 1959.)

Simply put, we need to return to waging war the way we have proven it works. We need to reemploy the moral positions, thinking and the strategies we used so effectively in WWII and the Cold War and abandon the failed policy of trying to win over the hearts and minds of undefeated people who hate us and whose religious beliefs tell them to kill us simply for existing.

It is the people who are the ones who hate. All of the polls and interviews posted in the past several years reveals the deep, entrenched hatred of America and the West held by the people of Arab societies.

Americans are easily placated by the media which frequently shows a few educated Muslims, usually students or other relatively liberal individuals who express some desire for a more Western life in their cultures. This complies with our sense of political correctness. These people are few and far between among the people as a whole. Most Muslims throughout Arabia, the Central Asian countries and Iran and Indonesia are medieval Muslims who comply completely with the teachings of their clerics and mullahs.

Throughout the Arab World and even into Europe and America Saudi money supports madrasses that overtly teach the inferiority of Jews, Westerners, and all other infidels while exalting the idea of Holy Jihad and Martyrdom. To die in Jihad is the dream of Muslim boys and young men (and increasingly women) throughout Islam. The same things are being taught by the Iranian theocracy.

The combination of great idleness and great wealth among Saudis along with their religious intolerance leads easily to grandiose ideas of self-importance and a vision of their ruling the world for Islam. It is not by accident that almost all of the 9/11 hijackers were Saudis. Although Bin Laden has suggested that only the team leader knew the missions were suicidal, all were willing accomplices of the crimes.

The Saudis were totally indifferent to the slaughter of Westerners, even 9/11, until Osama Bin Laden targeted the Saudi Royal Family for overthrow for allowing the Americans on Holy Ground during the first Iraq War. After the truck bombing of the Kobal Towers which killed around two hundred American servicemen the CIA and FBI received no cooperation from the Saudis in investigation of the murders and the Saudis killed those they thought were responsible without allowing us to question them. They cared nothing for the Americans who were keeping them safe and in power. To the Saudis, all others are merely their servants.

Chapter Four

Atomic explosions have devastated
Omaha, Nebraska and Denver Colorado within forty
minutes of each other resulting in the deaths
and injuries of over one million Americans.
Al-Qaeda, through the news organization,
Al-Jezeera, has demanded the surrender of the
United States of America and
submission to the rule of an Islamic Religious
Council
Otherwise more cities will be destroyed. When
this day comes, how do you want the President
to respond?

In 1965 Pakistan started creating atomic power at Parr, Rawalpindi.

During the 1970's and 1980's Pakistan acquires technology and material to produce atomic weapons while assuring the world with announcements and treaties that it has no intentions of creating such weapons. Technical support came primarily from China.

In 1984 President Reagan issues an empty threat of "grave consequences" if Pakistan enriches uranium past five per cent. Pakistan was not deterred and no "grave" action was ever taken.

In a 1986 interview President Zia of Pakistan said that they will have a nuclear bomb and that the world of Islam will have it with them.

In 1986 Dr. Abdull Qadeer Khan the "father" of Pakistan's nuclear program states that they have the bomb.

May 28, 1998 Pakistan detonates five atomic bombs.

In 2004 the CIA arranges a boarding of a Pakistani ship bound for Libya carrying centrifuges and nuclear technology. Moammar Kadafi, remembering the bombs America dropped on his house a few years before, came clean and agreed to give up all nuclear ambitions and "spill the beans" on Dr. Khan just so long as we didn't bomb him again.

In 2004 Dr. Kahn admits to widespread nuclear proliferation of materials and technology to several end buyers which had made him a multi-millionaire.

2004 President Mushareff pardons Dr. Kahn of violating Pakistan's pledge of non-nuclear proliferation, in large part because Kahn is beloved by the Pakistani people.

We have seen that when it comes to American concerns, threats, and bravado over nuclear proliferation, one thing is abundantly clear: the world only pays attention to action. We threatened, with zero effect, to stop the development of nuclear weapons by the following: India, Pakistan, Iran and North Korea. The latter has put forth mixed results as on the one hand Kim Jong Il has agreed to open up his nuclear facilities (in return for ransom) for inspection while at the very same time is providing nuclear technology to Syria...not a situation to be trusted.

We didn't bomb any of these countries and our threats meant absolutely nothing to the leaders or people of those countries. One country, when confronted with proof of trying to build atomic bombs, did understand and fear our threats, and that was Libya. This was because we dropped a bomb onto the house of the leader, Kadafi. He understood. And that was the end of that threat.

There is clearly no rationale to accept any assurances from Pakistan or any other of the Arab or Iranian governments. Currently Iran is trying to play the deception game with world atomic energy authorities with success. The world is allowing the radical Islamists the time and the circumstance to have delivered into their hands the weapons of mass death that will truly threaten and will almost certainly kill millions of Americans.

There have been numerous reported announcements and quotes from Al-Qaeda that they would use the bomb as soon as they could. They love the idea of killing tens of millions, even billions, of non-believers. Given the creativity that allowed nineteen guys armed with box cutters to kill 3000 people and do billions of dollars of damage on 9/11 we are really being suicidally stupid to fail to comprehend the threat they pose. (The death toll could've reached 50,000 had the Muslims attacked later in the day when the buildings would have been full...as was certainly their intent.) We should guard against being so filled with bravado that we don't want to admit the danger. We should proclaim to the world that we fear for our very lives and for our country and way of life to the extent that we conclude we are entitled to take whatever action we deem in our best interests and defense.

One nuclear power in the Middle East is Israel, It is popular to think that the whole issue is that Israel exists and the plight of the Palestinians is the cause of Muslim hatred. This is not true. Osama Bin Laden has urged the faithful to delay the issue of fighting for a free Palestine and concentrate on the more important enemies. These he names as first, heretics (those who work with the Americans, like President Mubarrak of Egypt), second, Shiites (about twenty per cent of Muslims), and then America. The Palestinians can wait.

All Arabs and Jews are Semites (excluding converts such as Ethiopian Jews). It is always amazing when ill-informed politicians proclaim the Arabs are anti-Semitic; they are anti-Jewish, not anti-Semitic since they are

Semites. They are all from the same tribal stock going back several thousands of years.

Chapter Five

Atomic explosions have devastated Philadelphia, Pennsylvania, and Phoenix, Arizona within forty minutes of each other resulting in the deaths and injuries of over one million Americans. Al-Qaeda, through the news organization, Al-Jezeera, has demanded the surrender of the United States of America and submission to the rule of an Islamic Religious Council

Otherwise more cities will be destroyed. When this day comes, how do you want the President to respond?

In 1942 Jimmy Doolittle led a squadron of fighter-bombers to retaliate against the Japanese for the attack on Pearl Harbor. This was the first counter-punch America offered the enemy. Due to the bombing technology of the day and the vast distance the planes had to fly even to get from a safe take-off point to Japan the raid was not very effective from an operational standpoint. More importantly it let the Japanese know they were vulnerable and it was a huge surprise to the people of Japan. It is especially noteworthy that the American President Franklin Roosevelt and the military commanders did not target a military base or naval shipyard. They wisely chose to bomb the city of Tokyo.

Throughout World War II, the Allies realized the imperative of destroying the enemy populations. Ultimately all of the cities of Germany were leveled with bombs or even more effectively, with incendiaries which literally turned the cities into infernos that burned almost everything. General Le May used these weapons to burn Japan to the ground. Only three major cities were spared. Kyoto was preserved because of its historical value and because of the anticipation the war would soon be won. The other two cities were Hiroshima and Nagasaki. The atomic bombs which leveled these cities, killing perhaps one hundred thousand people in a day, finally brought about the surrender of Japan.

Wars are waged by people against people. War material may certainly be a legitimate target for it prevents the enemy people from using it against us. It is proper to destroy an arms factory. In the case of the Arab and Iranian Muslims, the people are the arms factory. Enraged by self-induced hatred the Muslim people are the ones waging war on us. They are not ordered to do so by their governments, but by their religious leaders and the most influential of their followers. It is doubtful that there would be any terrorists if they were not encouraged by their families, mullahs, relatives, neighbors and their media. They live in a world where everything from the graffiti on the walls to the teachings in the madrasses, at home and in the mosques, are to prepare them for jihad, for holy war against the West and the vile infidels.

It is not any more likely that we will be successful winning the hearts and minds of the Muslims than we did with the Vietnamese. Any reliance on changing their minds with any influence we will have is suicidal. Only the mullahs would have any influence and the moderate mullahs are losing influence to the radical elements. The idea of war and religious righteousness will prove too powerful as it takes on a life of its own. Even Al-Qaeda has morphed into a self directed web of cells of individuals that, like a wild fire, moves as opportunity presents, without necessarily being directed from a central authority. That is not to say that there will not always be central direction and influence.

Given our complete victory in WWII that showed us how to defeat a huge, powerful and completely evil enemy, it is perplexing how we reached a moral position that we somehow deserve to be killed by an enemy. Presumably it is the aftermath of the Vietnam defeat. America was wracked with guilt, shame and a loss of confidence in the correctness of our efforts. This lasted for decades. Even today, we seem to think we owe our blood and our money to bring peace, democracy and prosperity to people who want to kill us, as is the situation in Iraq and Afghanistan. How did we ever get to such a national state of mind?

In fact it is our responsibility to do now what we did in WWII. We know who the enemy is. It is the Arabs and Iranians (not to leave out the other Muslim societies). During WWII we had no such ambiguity about what had to be done and how to do it. We had to kill the enemy army and navy combatants and we had to kill the general populations as well. This was necessary, appropriate, and moral.

When we waged war in the forties did we not kill snipers, entrenched machine gunners, young children, submarine crews, blind elderly in beds, fighter plane pilots, battleship crews, people who never wanted war at all, and bomber crews, to name just a few categories? Of course we did. It is war. It is them or us. After the war we found out what the Nazis did to the Jews, what the Japanese did to Chinese civilians, just as we see on videos from Iraq of Muslims cutting off the screaming heads of bound innocent non-combatant hostages in addition to the falling World Trade Towers. It is war and it is long past the time we American people accept it.

It is almost humorous to consider how we would have fared in the war with Germany and Japan and Italy if we thought our responsibility was to "save" the German people from the Nazis, the Italian people from the Fascists and the Japanese people from the Imperialists. What if we had thought that the right thing to do was to somehow capture the leadership and bring them to trial rather than wage war? The people of any population are inevitably responsible for what the group does. If the people lose control of the few who do the great depredations on others then they must bear the consequences. Of course there were Germans who in their hearts were totally against waging war, just as surely there were Japanese who were happy with their lives and wanted nothing to do with war and slavery. This did not make the taking of their lives along with the guilty a bad or immoral act. Collective guilt is truth, inescapable and proper.

Are Arab people more deserving of life than Germans? Do Iranian elderly more deserve to be spared than Japanese elderly? Of course not. We cannot allow a false morality to endanger our own lives when it is us who are the initial targets of war. If it was moral to wage total war against the people of Germany and Japan it is just as moral to wage total war against the Arabs and Iranians now.

The Presidential Oath of Office reads: "I, _____ do solemnly swear that I will faithfully execute the Office of President of the United States, and will to the best of my ability, preserve, protect and defend the

Constitution of the United States. (So help me God.)" The last phrase was first added by George Washington and most Presidents have continued to add that statement although it is not in the Constitution.

Nowhere in the job description of the Presidency is there a requirement to bring peace, democracy and prosperity to foreign people. We do not owe them our lives, our money or our prosperity. To offer help is one thing, but to risk our very lives and the life of our country and culture to people who don't deserve our help is senseless. Perhaps it is a product of the of the Texas Bible Belt mentality since President Johnson (Vietnam), President George H. W. Bush (Iraq I), his son President George W. Bush (Iraq II) all gravitated to this approach. Perhaps it is their desire to play "Texas Ranger to the World".

The Presidents Bush and President Carter are all devout Christians embracing a benevolent view of the world and desiring a world of a "thousand points of light" and a commitment to the idea that America and the free world would be best served by bringing democracy to Muslim countries and dealing with them as gently as possible.

When Saddam Hussein ordered an assassination hit on the former President Bush Sr., during a trip to Kuwait along with his wife and the wife of the second President Bush that was reason enough to invade Iraq and kill Saddam. Killing an American President clearly undermines the position of not only the office but puts a chill on all future Presidents in their efforts to promote the interests of the American people.

Our biggest problem of pure self-defense is that we have no deterrence to being attacked. We have taken the position that we can't kill the populations, which is the only sensible thing to do and the only strategy that has shown itself to work over a long experience of many wars and conflicts, both real and conceptual. A completely credible threat of total destruction might bring about the end of this threat, but it is doubtful.

A possible combination of actions and threats might work though and be worth trying. One scenario might be to destroy several major holy sites, like the city of Kofu in Iraq and Medina in Saudi Arabia to let the Muslims see we are serious and they will pay an unacceptable price. This along with the threat to destroy the holy Kabala in Mecca and all other holy sites (excluding Jerusalem) would deter them.

We found in the writings of the Devoted to Allah, the 9/11 warrior Mohamed Atta, how he cleansed himself before embarking on his holy mission of Jihad. He knew it was necessary for him to be completely clean, his skin, his hair, his clothes…all must be ready to be accepted into heaven after his act of devotion to Allah. He went to considerable lengths to clean himself, to pray, to prepare himself before he would present himself to Allah upon his Martyrdom. From this we can learn a great deal about those Devoted To Allah.

One deterrent would be to execute Muslim terrorists by dipping them into a vat of pig feces and urine and have them drown quickly and humanely, yet doing so in a manner that will deny them Heaven. They would be so obscenely unclean Allah would not allow them into Heaven.

To our founding fathers, "cruel and unusual" forms of execution would certainly have considered the electric chair, gas chambers, and lethal injections to be "unusual" to say the least. It is not improper for us to use a method of punishment that will deter crime, since that is one of the primary goals of capital punishment.

As a nation we are constantly reappraising our attitudes toward the enemy and how to treat them. Punishment, torture and executions are all being discussed for we are a country preoccupied with our self image and how we are perceived by the world.

After 9/11 occurred and we realized we could be attacked again, much discussion followed about the appropriateness of torture to gain information from the enemies we capture to prevent other acts of terror. Even very liberal proponents as Prof. Allen Dersherwits of Harvard University proclaimed that "of course" we would and should use torture if that would prevent mass death and destruction.

It is always a perplexing question why we will kill 100,000 people in a day in the conduct of war and still agonize over the treatment of a single prisoner.

We have had the luxury of conducting warfare for the last one hundred fifty years on the soil of others. However we conducted the war held little likelihood that America would be truly at risk... Only Japan actually attacked US bases. Germany, Italy, Korea, China, Vietnam, Panama, Granada...never did they attack us on other than the foreign field of conflict. Then 9/11.

We are now faced with the inevitable attack with weapons of mass destruction that can totally defeat us. Pakistan detonated five nuclear weapons simply as a show of force to India. Clearly many more rest in the arsenals of Islam. The President of Pakistan is a dictator friendly to the US, but generally not supported by the people of Pakistan who are extreme Islamists. From the madrasses in Lahore that teach radical Islam to children, to the violent, western provinces whose people openly support the Taliban all are an indication that Pakistan is a coup away from delivering nuclear weapons into the hands of those who would delight in using them to destroy us.

This threat is real and it must be eliminated. The answer is simple and clear. First we must obliterate the nuclear facilities of Pakistan. Our troops which thankfully are already on the ground in the middle of the oil fields should bc rcinforced and used to depopulate the areas of the oil facilities in Iraq, Iran, and the other Persian Gulf states.

Only by taking over the oil facilities can we both deprive the Muslims of the wealth of the oil revenue but also to secure it for our own use. We should consider the assets to belong to the citizens of the United States of America in compensation for the damage threatened and done to us by the Arab and Iranian Muslims, especially the Saudis.

As an aside, the American ownership of all the Middle Eastern oil would solve many of our most pressing economic challenges. All oil wealth should be administered by a separate entity much like the Federal Reserve Bank was established. All profits, after the costs of production, distribution, security and so on would be paid directly to each American citizen over the age of 18. The total would be somewhere between $1,500 to $3,000 a month to each citizen, after about one fourth to one third allocated to Social Security to provide health care and a comfortable retirement to all Americans.

Consider what if, during the 1930's while Germany was rearming and the world could read Hitler's book "Mien Kempt" where he proclaimed what he wanted to do, the nations of America, France and Great Britain had attacked Germany and destroyed the German war machine and killed hundreds of thousands of Germans? By attacking early on when the Versailles Treaty was violated and before the Germans had rearmed Hitler and Nazism could have been brought to an end.

So many millions of innocents would have been saved a gruesome death. Of course today, we would look back on such action as horrible, attacking innocent Germans and killing them for no good reason! The horrors of the European War and the Holocaust would never have happened and we would never know what would have been prevented.

Ask yourself, "Do you revere those veterans who fought so bravely and EFFECTIVELY in WWII?" They killed the enemy populations. They killed them by the millions. To protect and save the free world they came to terms with the morality of killing millions of people. Are we now better than they?

What we face now with the Muslims is the same only the numbers are larger.

CHAPTER SIX

THE WILL OF GOD

The United States launches nuclear attacks on all Muslim military and population centers throughout Arabia and Iran sending tens of thousands of grateful Muslims into Martyrdom

In all of human history one's entrance into heaven or otherwise more benevolent afterlife, from the times of ancient Egypt, China, the Middle East to the great religions of today: Hinduism, Judaism, Christianity, Buddhism, and the others, was determined not how one dies but how one lives. From the Egyptian Book of the Dead where we learn that 5000 years ago one's heart was put on a scale to be measured for goodness to be allowed into a better afterlife to the Christian belief today that one must live one's life in goodness, kindness and charity to others to enter into heaven. Only the Muslims determine the most attractive Heaven by not how one lived, but rather, how one died. Only those who die in Jihad, regardless of the quality of their past life get the finest Heaven of all, the Martyr's Heaven.

Ultimately, and ironically, the only way we will be able to bring together our disparate beliefs and deal effectively with the Nation of Islam is to embrace it. Only by truly appreciating and respecting their beliefs will we find the proper and necessary way to confront and defeat this greatest of all threats. At first this may seem strange, even foreign, to most Americans but it is the only way.

Finally, God's will is apparent and clear. The two greatest religions of humanity, one and one half billion Christians and one and one quarter billion Muslims are at last at a place of oneness. The only two great monotheistic religions have now become one in destiny and completion.

Christians are compelled by Jesus's teachings to love their enemies.

Muslims are compelled by Allah's Revelation to the Prophet Mohammad to defend Islam against all threats…to be a martyr, to die defending Islam. The presence of any other than Islam on earth threatens Islam, therefore any Muslim who dies at the hand of the Infidel is blessed with the greatest of all heavens, the Martyr's Heaven.

By truly and completely respecting Islam we can accept what they believe to be true and from this point of view we can save ourselves while we respect and even love the Islamists. It is clear that while all true believers of Islam go to heaven…if they are not foul when they die…it is also clear that the Martyrs have a very special heaven which only they enjoy. There seems to be some disparity between Islamic clerics about the details of this heaven but some things are beyond doubt.

Martyrs enjoy the 72 virgins, rivers of wine and honey, the finest food, and most of all, unending sex for all eternity. All a Muslim has to do is die fighting the infidel…which is us…and this most special of all heavens will be his or hers for all eternity. We must ask ourselves, "Who are we to offer disbelief of this tenet of Islam? Who are we to deprive the Muslims of this greatest of all gifts when it is within our power to deliver them this blessing?" If we are to truly accept and respect Islam must we not also accept and support their belief that these wonders will be theirs if only we kill them? Heaven has no exception to age, sex, or action of lack of it on the part of the Muslims, only that they die in Jihad against the infidel.

There was a case in Israel, where some Hamas killers in a drive-by shooting shot to death a young man waiting for a bus. As it turned out, the young man was not a Jew but a Muslim going to school. Hamas clerics promptly proclaimed him to be a martyr, even though he was shot to death by fellow Muslims and not, to our thinking, even close to being engaged in battle to fight the infidel at the time of his death.

It must be seen to die a martyr requires only that one be Muslim, and must die during Jihad. Since we are clearly in a state of Jihad, every Muslim who dies at least a violent death, or dies of disease even remotely connected to the Holy Jihad dies a Martyr. There is no reason to think this must be limited to warfare, but must also include famine, freezing, starvation, heat exhaustion, evaporation, radiation poisoning, and other causes of death. This young man must even now be basking in heaven, eating sweet meats, intoxicated with endless wine, and having infinite sexual pleasures with his 72 virgins. This must be better than going to class.

If we truly love the Muslims, if we want the best for them, if we truly respect their religious beliefs, if we do not allow our own religious intolerance to reject their devotion to their clear message from Allah as brought to them by the only Prophet of Allah, then we have a cause, indeed a duty, to elevate them all into a state of martyrdom as quickly as possible. We must understand the sense of urgency since with over one billion Muslims on earth, many are dieing of old age, other natural infirmities, and common accidents every day and for those poor souls, they will never experience the ultimate Heaven of the Martyr. We can prevent this needless absence of ultimate heaven by acting quickly to get these people into a state Martyrdom.

We in the West tend to associate killing people as essentially a bad thing. We have accepted it in generally unusual contexts such as war, capital punishment, self defense, and rarely even euthanasia and cannibalism when deemed appropriate. It is a challenge for us to accept it as other than at best a lesser of evils. Now we must consider it as a true blessing as we accept and respect Islam.

One of the biggest challenges is to understand and to grasp the difference in terms of time. We want death to be put off as long as possible and nothing is more important than making life last as long as possible. This brings us to the concept of eternity. We know mortal life is short term compared to eternity but in Western thought and religion these do not pose a conflict but for the Muslim it does. We do not comprehend eternity as do the Muslims. Keep in mind that Muslims contributed the Arabic numerals (better to work with in decimals than Roman numerals) and also the decimal system and the concept of zero, which the decimal system requires.

Many Westerners still have a problem with the concepts of eternity. A simple example is to think of dividing something by 3. In fractions one can always get an even one-third but not so in the more exact decimal mathematics. Imagine the simple problem of dividing 10 by 3. We get 3.33333333333333333333 to infinity. There is never, EVER an end to the threes. Ask yourself, do you think that SOME TIME it would have to end and come out even? If this is your conclusion, you are among those who really do not comprehend infinity or eternity.

This same state of mind makes it hard to accept that the Muslim Heaven lasts for all eternity and to risk losing the infinitely incredibly, uniquely wonderful Martyr's Heaven by wanting to live a few more mortal years would be the stupidest of all choices in life. Up until the moment of Martyrdom the hapless Muslim risks dying in a car crash, of disease, getting shot by another Muslim…all kinds of death could happen which would then deny the Muslim the Martyr's Heaven FOREVER!

For the first time in human history one people have the power to send to heaven the entire population of over one billion souls. Imagine what a tremendous honor and blessing this would be. Just for a moment, ask yourself, "If Muslims could and did bring about the ecstasy of the Rapture, who would condemn it?" What Christian would not welcome the second coming of Christ? If you were, at this very moment, ascending to be joined with Jesus by an act of the Muslims, would you not be grateful? Would you not be filled with love and gratitude that not only is your wait over, but that you were among the living able to experience this greatest of all soulful experiences? Now we all know the Muslims would never be willing to do something so wonderful for anyone else, but are we not more benevolent, more charitable, more kind, more willing to sacrifice ourselves for the good of others enough to create for them their path to Martyr's Heaven even while we protect our own?

The Christian duty of all of the faithful of Jesus, those who will truly love their enemy, is to take up the Cross and become Christian Soldiers marching unto War, Holy War. How else can so many do so much for so many others? Is not the gift of eternal special heaven worth giving?

Muslims cannot do the rest of us any favors by killing us and sending us to heaven as we all enjoy what ever afterlife as a result of how we lived, not how we die. We are much better off living a long life and dying of natural causes…something the Jihadists are fervently seeking to prevent.

Make no mistake…the Muslims will not like this and will reject this position for it denies them the domination of the earth for Islam. Since they will do and say anything to further the goal of an Islamic world, which they are quite willing to do by the sword, we must expect them to decry this proposal.. We should be willing to give the Muslims eternal heavenly bliss but we should not surrender the world to them.

Part of the wonder of God's Plan is that the Western mind values the individual over the group. We expect the institution to serve the benefit of the collective and singular individual. The individual serves the institution…nation, church, etc, only so long as it serves the individuals. This is the opposite of Islam (the word means "submission"…the individual submits totally to the "institution" of God). This conflict is resolved by the Western power delivering the individual into the Martyr's Heaven while not serving the institution of Islam for which the individual Muslim must sacrifice all to conquer the world for Islam. Here the Western ideal, the individual, achieves the most wonderful of all outcomes while preserving the submission of the individual Muslim to Islam. Remember, the Muslim does not attain the most desirable heaven by conquering the world for Islam, but by being killed in the Jihad by the Infidel. It is really a quite neat arrangement. Keep in mind the Muslims never go into to battle or Jihad proclaiming a desire for victory, they pray for Martyrdom!

It is also possible that the Muslim clerics are actually intending to goad the American superpower into giving every Muslim a Martyr's Heaven as a global version of "suicide by cop" where one cannot take one's own life so they threaten the police into shooting them to death. This is, after all, the only way all the Muslims could achieve this end result since they cannot be disingenuous to Allah.

If, after the initial period of sending Muslims into Martyrdom, the Nation of Islam determines that they do not want further entrants into Martyr's Heaven, the surviving Muslims must, at the minimum, be required to renounce certain tenets of Islam:

First, renounce Martyrdom.

Second, renounce Jihad or any other killing in the name of Islam.

Third, renounce all sense of Muslim superiority over non-Muslims.

Fourth, renounce the killing of apostates (those who leave Islam).

Fifth, renounce the domination of the world by Islam.

Sixth, accept a revised version of the Holy Koran from which all references to killing infidels is excised (this will shorten the Koran considerably). This would be not unlike the Christians accepting the King James' version of the Holy Bible which is widely accepted among Protestants.

Lastly, accept the transfer of oil assets and other wealth to the American citizens in compensation for the acts of Jihad waged against the American People.

Essentially, we must create a separation between church and state in the Muslim societies. It might be helpful to understand that in those societies that were the precursors of Islam there were no laws of any kind. What differentiated the tribes of Israel from the other Semitic tribes was that the Israelites developed laws from Moses. It is amazing to us coming from a Judeo-Christian past that a people had no law to guide them.. The Jews had laws for thousands of years but not the other tribes until Mohammed gave them the Koran as the basis for law. Those societies now believe that the only law is that of the Koran. They must be taught that secular governments can adequately establish law separate from religion. This will not happen without destroying radical Islam which is the product of the Saudi Royals and the Wahabi sect of teachings.

It is the Saudis that are fueling the fire of Islam. It is they who are printing the school books teaching the children to hate the infidel and to kill us and to desire Martyrdom. They are funding the schools not only throughout Islamic countries but all over Europe and even here in the United States. These are the lessons being taught to Muslims here in American Islamist schools and mosques. They teach them to want to be killers. The Saudi Royal Family wants to rule the world.

When it comes to fighting an idea, like Radical Islam, it has been said that one cannot kill an idea. This may be but it is moot since the same end result can be achieved by killing all the brains that choose to think the idea.

There is only one God. The God of Abraham is the God of Ishmael is the God of Moses is the God of Jesus is the God of Mohammad. Christians accept that the Jews are the Chosen Ones of God, and none but Jews will ever occupy that position. Similarly, God has provided Christians Jesus to be their personal Savior. So too has God made a separate arrangement for Muslims, as revealed in the Holy Koran.

As Christians know, Israel must regain its total dominance of the land God gave to them to usher in the return of Christ which in turn will bring total salvation to all True Christians. Although this means the end of Jewry as all will become Christians, this is God's will. By sending all the Muslims into Martyrdom Christians give them the best of all possible outcomes, the one meant for them alone by Allah, while at the same time ringing forth the return of Christ and the Rapture.

Many believe that God works in mysterious ways and that God helps those who help themselves. The confluence of these powerful energies, faiths, and billions of souls cannot be by happenstance. Only the Hand of God propels the true believers into this holy war with the outcome to bring to Muslims and Christians their most cherished and longed for conclusion to the faithful. At long last Jews, Muslims and Christians will have religious common ground and will be able to love and find the greatest gratitude in their hearts for each other, for without each one none would be able to reach the zenith of their respective beliefs. Such is the love of God to all the faithful.

CHAPTER SEVEN

GLOBAL WARMING

Undeniably the earth is experiencing a major climate change that if left unchecked will wreck vast destruction upon the planet, especially the Northern Hemisphere. This is due to the "greenhouse" effect that reflects the sun's heat back onto the earth and as the average temperatures rise, the polar ice caps will melt. First will be global inundation of all the coastal areas resulting in great calamity. Then this will change the ocean currents as massive amounts of fresh water will affect the salinity of the oceans, especially the North Atlantic, changing the characteristics of the Gulf Stream and this would bring about a new ice age that will destroy most of the heavily populated areas of the planet, destroying huge numbers of humans and decimating innumerable animal and plant species.

Much has been written about global climate change and it is unnecessary for a repeat of it here. Most people are convinced of the correctness of the claims by many scientists about this phenomenon.

There is some dissent about the cause of the change, whether it is a continuation of the natural warming of the planet which has been going on since the peak of the last ice age and or to what extent human activity has caused or hastened this change. Certainly, humans have contributed to climate change down through the ages. Much of the argument today centers around the increase of carbon in the atmosphere due to human activity especially in the industrialized and developing nations.

Consider that humans, according to theories of evolution and the fossil record, have been around for about 150,000 years as fire making peoples. We don't know when humans first discovered fire but we have evidence of it going back long before the last ice age. Humans have endured a lot of ice and cold in the past.

Consider that early humans would have started what would be unimaginable forest and grassland fires by accident. We have the image of cavemen sitting around a fire keeping warm and cooking their hunks of meat but think about the little caveboys playing with fire.

Once those fires would have gotten started there were no fire crews with water dropping planes to help contain them…they would have burned vast swaths of ancient land until they burned themselves out. Huge amounts of carbon would have been thrown into the air and over time might well have ushered in the last ice age for that matter.

We also know that volcanoes contribute amounts of ash and smoke that are enough to provide a short term effect upon the weather world wide offering lower temperatures for months and even years afterward. Climate change has always been part of the history of the earth and that will not stop.

Our concern today is more rooted in fear for the immense populations, especially in North America and Western Europe, which would be very adversely affected by the flooding of low lying areas. Most populations in these countries live along the coasts and are vulnerable to catastrophic flooding. The cost and disruption foreseen over even the next generation is disturbing enough that nations and people all over the world are striving to find a way to solve this rather short term problem and at least spread out over many centuries the effects of the climate changes we are seeing accelerating.

Disaster is also seen in the effect of climate change on farming and water supplies even in the best equipped and most prosperous nations. Whole cities may run out of water, droughts might mean the end of entire farm belts upon which hundreds of millions depend upon for food. The water temperatures in the oceans are changing so fast that many species of food fish are not able to adapt and large fisheries may be destroyed leaving millions more humans with inadequate sources of food. Fears of economic disasters in the short term are keeping the most polluting nations from reaching accords that might help alleviate the crises.

Many scientists, seeing the statistical curve of the progress of the change, are reaching the conclusion that merely curbing greenhouse gases, reducing the carbon footprint of people and industries is simply too little, too late. Whatever is done, even under the most wildly optimistic scenarios, will not end the problem foreseen by the melting polar icecaps, which are already too far gone to reverse the greater reflected heat off of what was once ice and is now heat absorbing land and water.

The problem is impending catastrophe for humanity and untold species of other animals and plants. While life on earth will certainly continue, we want the earth we know to continue. To achieve this end we must do something immediate and of enough impact to alter the climate in a profound and immense manner.

Natural history and our own historical record indicate that the only natural phenomenon known to achieve this in even a small way are erupting volcanoes. Only such incredible amounts of pollutants thrown up into the upper atmosphere can cover the earth to the extent of blocking out enough sunlight to rapidly cool the earth and return large amounts of snow and ice to the polar latitudes can have enough of an effect to offer any chance of success.

Try as we might we have no technology that would even remotely offer the ability to create volcanoes to erupt and save the planet from this impending disaster. We do have access to one extremely powerful tool that would accomplish this: Nuclear bombs.

Viewed alone, the idea of detonating large numbers of thermonuclear explosions has been studied since the fifties in anticipation of their possible use in war against the Soviets or the Chinese. The effects of radiation, immediate and residual, are well known. The amounts of material blasted into the atmosphere, the altitude and the resultant travel over the face of the earth due to the prevailing wind patterns is well documented and very well understood.

The effects of radiation has been known and understood for decades as has the effects down the years upon soil and land both around the immediate blast zone and going out for many miles. The yields in terms of kiloton and megaton are again well researched and documented. We have seen that such explosions do not contaminate the earth even under the blast itself to the extent that the land is uninhabitable. The now thriving cities of Nagasaki and Hiroshima, Japan attest to this.

Also reportedly, total nuclear war as once threatened by the US/Soviet Cold War confrontation would bring about a Nuclear Winter that would usher in a new ice age. This would be caused by the huge amount of dust and debris blasted into the heavens by all the nuclear detonations…cooling the earth and ushering in extreme cold as the earth would not be warmed by the sun.

If an extreme global warming will bring another ice age, and total nuclear war with thousands of atomic blasts would also bring about a new ice age, it is apparent what path we should take.

Creating a Nuclear Autumn would be the best way of creating a global climate balance. Detonating a smaller number of nuclear bombs, especially over one of the hottest areas on earth, Northern Africa, Iraq, Iran, and the Middle East would provide this necessary cooling effect. This would bring about incalculable benefits for all of mankind and nature and even for those who are not Jew, Christians, and Muslims.

Scientists have considered this general option for some time although the common approach is to inject into the atmosphere a severe pollutant, like one of the several sulfur compounds. This would be a great source of pollution but this is deemed worth the risk given the tremendous threat to the entire planet's ecosystems of global warming. It seems a shame to pollute the earth so severely and repeatedly since this stuff would precipitate out of the sky as acid rain, and needs to be constantly replenished. This would accomplish nothing more than preventing or mitigating global warming while if we detonate atomic bombs over the Nation of Islam we could achieve global peace, provide all those Muslims with the most wonderful blissful Heaven Allah can provide, stabilize the earth's climate, preserve countless species, and eventually usher in the Rapture.

All this has come to pass as God's working his will for not until now in all of human history have all these powers been in position to achieve this final goodness. God does work in mysterious ways

CHAPTER EIGHT

RECAP

There is only one God.

The knowledge of, and religion based on, the One God comes from the Semitic tribes of the Middle East, including the Israelites and the Arabs.

The Jews are God's Chosen People. God gave to these people a system of law, unique to the region.

This law is reaffirmed by Moses with God's gift of the Ten Commandments to the Jewish people.

Some Jews accepted Jesus of Nazareth as the Messiah and God gave to those people and all future Christians Jesus Christ as a personal Savior.

Christianity provides a life-everlasting or a heaven after death. This Heaven lasts for all eternity for those admitted to it. It is an ascetic concept of Heaven devoid of mortal pleasures (other than the overwhelming feeling of Love).

The Holy Bible is formalized by a council of bishops three centuries after Christ's death. All writings in the Bible are considered equally important and sacred.

Four centuries later God, Allah (from the Aramaic language Al-Lah, meaning "The God") sends to his Prophet Mohammed through the Archangel Gabriel, speaking to Mohammed in Arabic, the Revelations which over twenty some years constitute the Holy Koran.

This new religious requirement is Islam which means "submission".

The Holy Koran is not a collection of writings or revelations from different authors but are the continuous, in order, revelations of the will of Allah to his Prophet.

There are no contradictions in the Koran; where first there is a command from Allah for Mohammed and his faithful to respect and show mercy to non Muslims then later commands to kill all non-believers, all infidels, this simply reflects a change in the message from Allah.

Allah commands Muslims to conquer the world for Islam, by conversion if possible, by force…by slaughter of the non-believers…if necessary.

Every Muslim must defend Islam with his or her life, if required to do so by any threat to Islam.

The existence of any non-believers on the earth threatens Islam.

Any Muslim who abandons Islam threatens Islam and must be killed. (Thus, it is the duty of

Muslims to go to other nations and proselytize those people to convert to Islam but it is forbidden upon pain of death for other religions to proselytize Muslims for this condemns the Muslim to death if they convert.)

With Muslims demanding the right to convert others everywhere in the world and by forbidding other religions from converting Muslims they would eventually succeed in converting the world to Islam.

The Nation of Islam is medieval in nature. The majority of Muslims around the world are not well educated, are poor, and not traveled. They occupy the poorest and most isolated areas of the earth. Their way of regarding religion and authority and the proper conduct of daily life is religion dominated. This is essentially the same as medieval Christians of a millennium ago.

Medievalism essentially provides for endless killing of non adherents to the religion, education is only for the promotion of the religious life and people are intractable in their religious positions and are ready to kill to defend the religion.

All Muslims submit every consideration to Islam and therefore may enjoy many things forbidden on earth once in Heaven. This includes alcohol which Muslims are forbidden as mortals but is one of the most important gifts of Heaven, especially Martyr's Heaven.

To reward those who sacrifice all for Islam, meaning they die for Islam in the battle to conquer the world for Islam, Allah provides for the Martyr, or Shahid, the single most wonderful blessing from Allah…the Martyr's Heaven.

God provides for completely unique "deals" or "arrangements" for His believers. For the Jews, they are His Chosen People, (still awaiting their Messiah). For Christians, He provides His only Son, Jesus Christ, to be their Personal Savior and an eternal Heaven, sitting at the right hand of God Almighty. For the Muslims, who are not ascetics and do not care about singing Angels or knowing God, he gives them what they treasure most, a tactile, touchy-feely, real Heaven as a Garden of Delights.

Only for the Martyr, those who are killed in Holy Jihad, does Allah provide Paradise. This Paradise is the most perfect Heaven for Muslims; a lush garden under a bright sun, shining upon the myriad sweet smelling flowers, soft shade and cool breezes wafting under a bright blue sky, flowing cool streams of pure water, the magical appearance of the finest foods whenever the Martyr is hungry, springs bringing forth the finest wines so the Martyr may be intoxicated forever. Most of all the 72 virgins, who after sex, become virgins again and again, for the Martyr's unending sexual ecstasy. This lasts for all eternity…forever and ever and ever!

Both Christianity and Islam offer life after death and since Christianity offers a Heaven devoid of sex and Islam offers a Heaven with unending sex, few people, especially young adults world wide, when offered the choice between Christianity and Islam, more convert to Islam if what they want is life after death.

If a Muslim is not killed in Jihad he/she loses forever the opportunity to become Shahid, or Martyr. This regrettable fate is that of most hapless Muslims who end up dying of accident, illness, or old age. Such

Muslims might get the lesser heaven for all eternity, and then, maybe not even that.

Being killed in Holy Jihad is the most wonderful outcome for mortal Muslims, the highest reward from Allah shall be theirs alone.

Islamists are demonstrating increasing success in converting non-Muslims from Western Europe and turning them into suicidal martyrs with the promise of Martyr's Heaven. This is happening even in Great Britain, France and Germany according to news reports.

What has been interpreted as mainly a male oriented Paradise is now being increasingly couched as a female oriented Paradise as we see from increasing numbers of female suicidal martyrs.

Christians are compelled by the teachings of Christ to love their enemies.

There is no greater gift of love than to bestow the greatest blessing from the One God upon a Muslim by a Christian.

The greatest blessing is to kill the Muslim in Holy Jihad thereby transporting the Muslim into Martyr's Heaven for all eternity.

All non-believers in Islam are the targets in Holy Jihad as has been proclaimed by innumerable Muslim clerics, mullahs, and Ayatollahs. This applies specifically to Americans.

Christians are compelled by the teachings of Christ to love their enemy.

Westerners in general are compelled by Political Correctness to respect Islam. All expressed demands of Muslims are considered sacred and compelled by their religion which must be respected even though their religion plainly states the aim of the religion is to kill all Westerners.

Islam is the only religion which compels its adherents to kill all other people who are not of the same religion.

We are at war with the Nation of Islam, not with the essentially meaningless secular states.

The Holy Koran compels Muslims to lie and deceive the Infidels as a means to conquer the world for Islam.

Muslims know the Koran says first to respect Christians and Jews and then later to kill all infidels, which includes Christians and Jews. The Muslims knowingly lie to infidels when they point to the first part of the Koran when they want to propagandize the Westerners.

The most extreme of the Islamists are the Saudis which as an example, whenever they are persuaded to attend any state function or international meeting it is understood that the Saudi emissaries will not shake hands with the Israeli delegates because the Saudis consider the Jews to be loathsome, foul, and unclean, like a pig. This is always tolerated by the American government who almost always is the party promoting the meeting.

America and the West do not have a deterrent to attack, conventional or nuclear, from Muslims who want to die attacking us anyway.

Muslims have nuclear weapons now (Pakistan) and soon more parts of the Nation of Islam will have them (Iran).

Trying to end the threat by the Nation of Islam by "winning their hearts and minds" will not be any more successful in this conflict than it was against Vietnam. (Defining politico/military success in how many schools, roads and medical clinics are built in Muslim countries like Iraq and Afghanistan will be as effective as it was in Vietnam. One might ask how much such construction contributed to our success in Vietnam as the Communists took over Saigon and Americans were fleeing for their lives from what is now Ho Chi Minh City.)

America and the West have basically two ways to successfully defend against the attacking Muslims: first to hate Muslims and kill them as we did the Germans and Japanese in World War II, or, second to love the Muslims and send them to Martyr's Heaven as a blessing.

If the first option is chosen, then total all-out war would be waged killing the Muslim populations until they surrender and abandon Islamic positions that are murderous to non-Muslims including revising the Koran.

If the second option is chosen, then the effort is to transport all Muslims into a state of Martyrdom with all the attendant blessings for all eternity.

Either choice will result in hundreds of millions of Muslims getting into Martyr's Heaven and the salvation of the Western World. This can be accomplished with hate or love.

With all the Muslims in the Middle East in Martyr's Heaven the Israelis will fulfill the requirement of regaining all the land given to them by God.

With the Israeli redominance of the Holy Land the time will be prepared for the Second Coming of Christ.

The Rapture can begin after this has been accomplished.

Global Warming is the single most horrible thing to threaten the human population of the earth. If you are in doubt about this Al Gore will eagerly explain it to you.

Global Warming also threatens with extinction many other species of animals and plants.

While the argument continues as to the degree of fault of humans, especially those in industrialized and developing economies, there is general world-wide acceptance of the problem.

The general consensus of the scientific community is that the cause of the climate change is the warming of the earth due to atmospheric change resulting in the "greenhouse effect".

Most scientists are now convinced that under the best possible scenarios the current momentum of warming will continue for decades and will result in the melting of the polar ice caps and the flooding of the coastal population areas on earth which include some of the densest population areas and cities.

There is no way to stop this by conservation alone.

The only short-term power we know of to cool the surface of the earth is to put into the atmosphere a substantial amount of pollution which will block the necessary amount of sunlight and thereby reduce the solar heat build-up.

Volcanoes are the only means we have seen that have recorded effects of actually cooling the earth.

We have no way of inducing volcanoes to erupt.

The only tools we possess that will actually cast enough debris into the atmosphere are atomic and thermonuclear surface detonations.

The detonations of numerous atomic bombs over the hottest areas of the earth, the Middle, North Africa, Iran, and Indonesia would achieve the necessary cooling to reverse the warming trend and return the necessary balance to the earth's climate.

This should be done immediately before the glaciers melt any further for they will not be replaced quickly by cooling the earth to the degree we need to not freeze ourselves.

Additionally time is of the essence before the Muslims would use such weapons to slaughter us, which, unlike the Muslims being transported into Martyr's Heaven, does us no good at all.

Speed is also necessary out of consideration of the Muslims who are dieing every hour of every day of accidents, illness and old age thereby losing forever the greatest blessing ever made available to any mortals ever to walk the earth…Martyr's Heaven. They lose this FOREVER!

Muslims must be expected to counter this proposal with a lie, most likely that merely being killed in Holy Jihad by the Infidel would not automatically provide the Muslim with Martyr's Heaven. This would be totally in keeping with the demand of the Koran to lie to the Infidel to defend Islam and strive to conquer the world for Islam.

God, Allah, has created a means for all his children to achieve the long standing goals of Heaven by bringing about the necessary technology (nuclear bombs) the blessing of Martyr's Heaven, the Rapture for Christians and Jews, and lastly the saving of the planet from global warming.

It was to the Jews God gave the insight, the conceptualization, the actualization (along with American resources) to create the atomic bomb. The revelation was only to Albert Einstein, J. Robert Oppenheimer (leader of the Manhattan Project and father of the atomic bomb), Edward Teller, father of the hydrogen bomb, and the other Jewish physicists. Not a coincidence.

God truly works in mysterious ways.

Thank you for reading this book. Now please read it again before making final judgments about it.